YOUR KNOWLEDGE HAS VALUE

Ankit Ponkia

A Simple Design and Analysis of Coaxial Fed Annular Ring Microstrip Patch Antenna For Wireless Communication Systems

GRIN Verlag

Bibliografische Information der Deutschen Nationalbibliothek:

Die Deutsche Bibliothek verzeichnet diese Publikation in der Deutschen National-
bibliografie; detaillierte bibliografische Daten sind im Internet über http://dnb.d-
nb.de/ abrufbar.

Imprint:

Copyright © 2014 GRIN Verlag GmbH
Druck und Bindung: Books on Demand GmbH, Norderstedt Germany
ISBN: 978-3-656-68698-9

This book at GRIN:

http://www.grin.com/en/e-book/275071/a-simple-design-and-analysis-of-coaxial-
fed-annular-ring-microstrip-patch

A Simple Design and Analysis of Coaxial Fed Annular Ring Microstrip Patch Antenna For Wireless Communication Systems

Ankit V. Ponkia

Asst. Prof. Dept. of Electronics & Communication Engineering, Shantilal Shah Government Engineering College, Sidsar Campus, Bhavnagar-340460, Gujarat, India

Abstract

In this paper design and analysis of annular or circular ring microstrip patch antenns and the fundamental terms related to design aspects and study of antenna is presented. Like many available variations of microstrip patch geometries annular or circular ring has received considerable attention due to its broadband nature when operated in TM_{12} mode and has smaller circular counterparts when it is operated in its fundamental mode TM_{11}. In this article mathematical analysis of annular ring patch antenna with design is presented and studied. The designed antenna operates at 2.4 GHz resonant frequency so can be used in ISM (Industrial, Scientific and Medical) band wireless applications. The proposed antenna shows good return loss, VSWR as depicted in the graphs.

Keywords: Microstrip Patch Antennas (MSAs), Circular Ring Microstrip Patch Antenna (CRMSA), Annular Ring Microstrip Patch Antenna (ARMSA), Circular Polarization, Voltage Standing Wave Ratio (VSWR)

1. INTRODUCTION

Microstrip patch antenna (MSA) consists of metallic or conducting strip on dielectric substrate covered by ground plane on other side and patch radiates fringing fields around edges [1]. For miniaturization of communication equipment compact microstrip patch antennas (MSAs) have been much attracted due to its light weight and light volume and better performance. However, microstrip patch antennas have limitation of narrow bandwidth which limits its application in practice [2] but there are so many techniques available to improve the performance of patch antennas. The geometries of patch antenna like rectangular, circular disk, angular etc. are available and their variation such as rectangular ring, semi circular disk, annular ring are used for some wireless and communication applications. Also as an alternative to regular or standard rectangular and circular patch antenna rectangular ring and circular ring or annular ring patch antenna are widely used because of their broadband nature and these configuration are smaller in size as compared to prior configurations so size will be less as compared to its rectangular or circular counterparts [3]. These antennas are low-profile, simple and inexpensive to manufacture using printed circuit technology [4-7].

The annular ring structure is a good resonator for TM_{1m} modes (where m is odd) (with very little radiation), and it will good radiator for TM_{1m} modes (where m is even) [7].

There are several good features are associated with this patch configuration viz. when operated in its lowest mode size of this antenna is small then circular patch antenna so in array application size of overall antenna will be quite small. Secondly it is possible to combine annular ring with second microstrip element and thirdly the separation of the modes can be controlled by ratio of outer to inner radii and by operating in one of the higher-order broadside modes, i.e.

TM_{12} the impedance bandwidth is several times larger than is achievable in other patches (rectangular or circular etc. patch configurations) of comparable dielectric thickness [8]. Figure 1 and figure 2 depict proposed geometry of annular ring patch antenna structure. The outer and inner radii are a and b respectively as shown in figure.

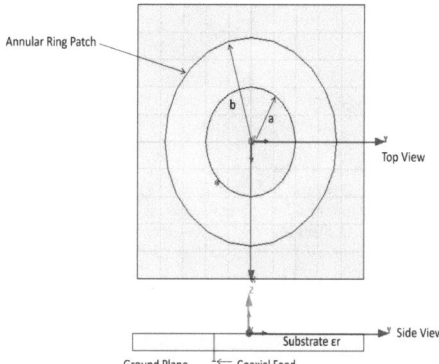

Figure 1 : Proposed Antenna Geometry: Antenna Dimensions $a = 9.75\ mm$ $b = 19.5\ mm$ Substrate Height $h = 3.125\ mm$ Patch thickness $t_p = 0.02\ mm$

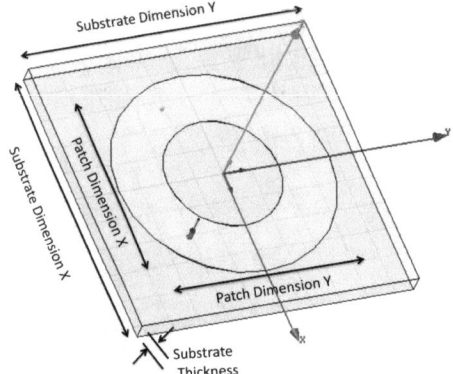

Figure 2 : Proposed Antenna Geometry Tilted View Substrate Dimension $(X \, x \, Y) = 50 \, x \, 50 \, mm$ Patch Dimension $(X \, x \, Y) = 39 \, x \, 39 \, mm$

2. ANTENNA DESIGN AND SIMULATION RESULTS

The resonant frequency of annular ring microstrip patch antenna (ARMSA) is always smaller then circular micro strip patch antenna (CMSA) [2]. The resonant frequency of ARMSA is given by [8],

$$f_{nm} = \frac{X_{nm}\,c}{2\pi b\sqrt{\epsilon_r}} \quad (1)$$

Where c is the velocity of light ϵ_r is dielectric constant of substrate, and X_{nm} represents the roots of the equation,

$$Jn'(CX_{nm})Y_n'(X_{nm}) - Jn'(X_{nm})Y_n'(CX_{nm}) = 0 \quad (2)$$

$J_n(x) \, and \, Y_n(x)$ are the Bessel Functions of the first and second kind order n, respectively and $C = b/a$, a is inner radius of patch b is outer radius of patch. Also, $X_{nm} = k_{nm}a$. In equation 1 effect of fringing field is not considered if we consider the fringing field the equation can be rewritten as below,

$$f_{nm} = \frac{X_{nm}\,c}{2\pi b\sqrt{\epsilon_{re}}} \quad (3)$$

Where ϵ_{re} is effective dielectric constant of substrate [9]. It can be determined [Schneider et al. 1969] by,

$$\epsilon_{re} = \frac{1}{2}(\epsilon_r + 1) + \frac{1}{2}(\epsilon_r - 1)(1 + \frac{10t}{w})^{-1/2} \quad (4)$$

Where

$$W = b - a \quad (5)$$

b=outer radii of the ring
a=inner radii of the ring and
h= thickness of the dielectric.

To account for the fringing fields along the curved edges of the ring, it has been suggested that the outer and inner radii be modified according to,

$$b_e = b + \frac{1}{2}(W_e(f) - W) \quad (6)$$

$$a_e = a - \frac{1}{2}(W_e(f) - W) \quad (7)$$

Where,

$$W_e(f) = W + \frac{W_e(0) - W}{1 + (f/f_p)^2} \quad (8)$$

$$W_e(0) = \frac{120\pi t}{z_0\sqrt{\epsilon_{re}}} \quad (9)$$

$$f_p = \frac{z_0}{2\mu_0 t} \quad (10)$$

Where,
μ_0 is the permeability and z_0 is the quasi-static characteristic impedance of a microstrip line of width W.

A pair of empirical formulas for the modified radii, sufficient for many engineering purposes, are given by [10],

$$a_e = a - (\frac{3}{4})t \quad (11)$$

$$b_e = b - (\frac{3}{4})t \quad (12)$$

The above model gives reasonably accurate results as long as W_e is less than the mean diameter of the ring, i.e. $(a + b)$. For the given values of a and b, a_e and b_e can be calculated. Then the characteristic equation is solved by replacing a and b by a_e and b_e. After solving the characteristic equation for k_{nm}, the resonant frequencies are- determined from equation (1) and equation (3).

In this article proposed antenna geometry is mounted on Roger RT/Duriod 5880 substrate material (dielectric constant $\epsilon_r = 2.2$ and loss tangent $\tan\delta = 0.009$) with thickness/height (h) of 3.125 mm.

2

Outer radius to inner radius ratio $C\ (= b/a)$ of the patch is chosen as 2 with inner radius (a) = 9.75 mm and outer radius (b) = 19.5 mm.

The simulation results of proposed antennas are performed by Ansoft HFSS^{TM}.HFSS stands for High Frequency Structure Simulator. It is full-wave electromagnetic (EM) simulator used to analyze 3D volumetric models and high speed, high-frequency component designs [11].

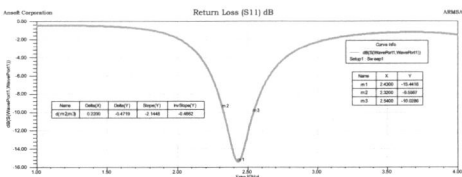

Figure 3 Return Loss (S$_{11}$) vs Frequency Plot

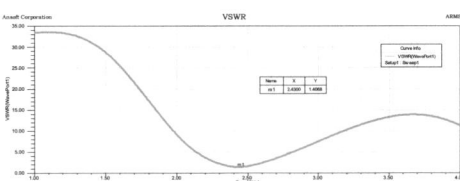

Figure 4 VSWR Plot

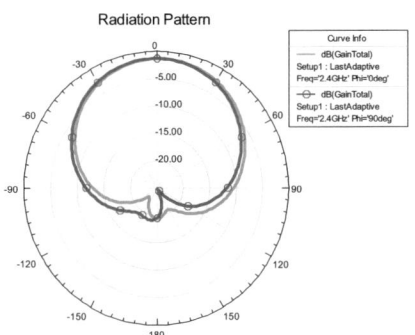

Figure 5 Radiation Pattern Plot at $\varphi = 0°\ deg\ (E-$ plane and $\varphi=90°\ deg(H-plane)$

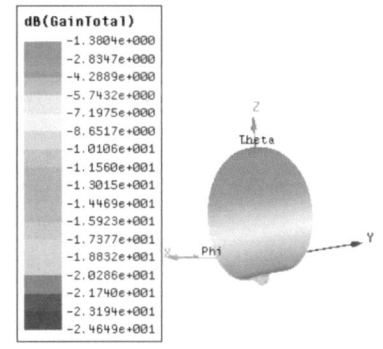

Figure 6 3D Polar Plot

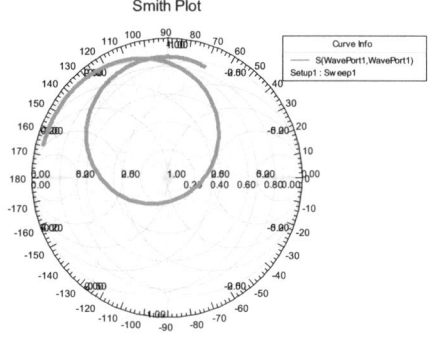

Figure 7 Smith Plot

The simulation results of antenna are shown in figure 3 to figure 9.Return loss at 2.4 GHz resonant frequency was obtained as -15 dB with VSWR of 1.4 and corresponding return loss bandwidth of 220 MHz (9.05 %) as shown in figure 3 and 4. Figure 5 shows antenna radiation pattern plot at $\varphi = 0°\ deg\ (E-plane)$ and $\varphi = 90°\ deg(H-plane)$.

Figure 6 and 7 shows 3D polar plot and smith plot respectively.

E-field distribution in patch and mesh refinement in patch is shown in figure 8 and 9 repectively.

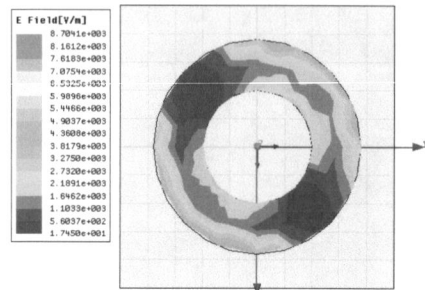

Figure 8 E-Field Distribution in Annular Patch

QUANTITY	VALUE (UNIT)
Max U	0.0559 (W/sr)
Peak Directivity	0.79086
Peak gain	0.72804
Peak realized gain	0.70248
Radiated power	0.88825 (W)
Accepted power	0.96489 (W)
Incident power	1 (W)
Radiation efficiency	0.92057
Front to back ratio	64.582

Table 2 Antenna Parameters at 2.4 GHz

3. CONCLUSION

In this paper annular ring or circular ring patch antenna design is presented with mathematical analysis. The proposed antenna geometry operates at resonance frequency of 2.4 GHz which lies in ISM band so this antenna can be used at various wireless applications in ISM band. Good retune loss and VSWR was obtained with compact design and simple structure.

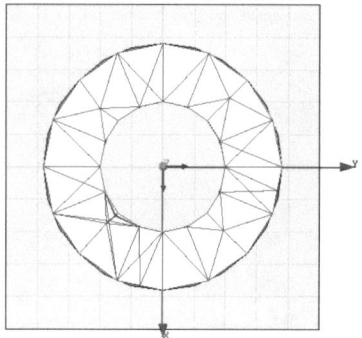

Figure 9 Mesh Refinement Plot

4. REFERENCES

1. Milligan T.A. Modern Antenna Design. *United States of America.:Mcgraw-Hill* Inc.1985
2. Kumar G, Ray K.P. Broadband Microstrip Antennas. Boston, London: *Artech House Antennas and Propagation Library*; 2003.
3. Kokotoff D.M, Aberle J.T and Waterhouse R.B .Rigorous Analysis of Probe-Fed Printed Annular Ring Antennas. *IEEE Transaction on Antenna and Propagation*, 47(2), 384-388, 1999.
4. Balanis C.A. Antenna Theory: Analysis and Design. 2nd ed., New York: *John Wiley & Sons*, Inc.; 1997.
5. Wong K.L. Compact and Broadband Microstrip Antennas. 1st ed., New York: *A Wiley-Interscience Publication, John Wiley & Sons, Inc.,* 2002.
6. Bancroft R. Micro Strip Antenna Design. *Noble Publishing.*2004.
7. Bahl I.J and Bhartia P. Microstrip Antennas.Dedham, MA: *Artech House.*1980.
8. James J.R and Hall P.S, Handbook of Microstrip Antennas, Vol. 1, London: *Peter Peregrinus Ltd.,* 1989.

Frequency (GHz)	Return Loss S_{11} (dB)	VSWR	Return Loss Bandwidth
2.4	-15.44	1.4	9.05 %

Table 1 Simulation Results of proposed antenna configuration shown in figure 1

9. Schneider, M.V. Microstrip lines for microwave integrated circuits. *Bell Syst. Tech. J.*48, 1421-1444; 1969.

10. WU Y.S and Rosenbaum F.J.Mode chart for microstrip ring resonators. *IEEE Trans., MIT-21*, 487-489; 1973.

11. Ansoft High Frequency Structure Simulator v11 User's Guide, 2005 *Ansoft Corporation.*